THE STORY OF HAY

Geoffrey Patterson

FARMING PRESS

My grateful thanks are due to Mr. J. R. Shawler and Mr. J. Carter at the Museum of East Anglian Life for their help with technical information about early machinery.

STORY OF HAY

Copyright © Geoffrey Patterson 1982

ISBN 0 85236 324 9

A catalogue record for this book is available from the British Library

First published in hardback by André Deutsch Limited
105-106 Great Russell Street, London WC1B 3LJ
Second impression 1983
Third impression 1987

Published in paperback 1996
by Farming Press Books, Miller Freeman Professional Ltd
Wharfedale Road, Ipswich IP1 4LG, United Kingdom

Distributed in North America by
Diamond Farm Enterprises, Box 537, Alexandria Bay, NY 13607, USA

Printed and bound by Proost, Turnhout, Belgium

Long grass that has been cut and dried is called hay. It is a nourishing food for cattle, and farmers need it to feed their animals during the winter months. Some of the grasses and plants that make good hay are shown on the opposite page.

For hundreds of years haymaking was done by hand using simple wood and iron tools. Then, in the early years of the nineteenth century, when inventors began to make machines for factories, they realised that machines could also be designed to do jobs on the farm. Farmers had always used horses for heavy work, now they found they could use them to draw machines which did the work of six or seven reapers in half the time. As machine design improved generally, so did the design of farm machinery. Today, haymaking from cutting to storing can be done in a few precious days of fine weather.

Italian Rye Grass

Cocksfoot

Timothy Grass

Perennial Rye Grass

Meadow Fescue

Red Clover

Ribwort Plantain

Lucerne

Burnet

wooden handle

reaping hooks

pick thanks

The reaper often used a hooked stick called a 'pick thank' to straighten the grass before cutting it. Although the reaping hook and pick thank were hard to use, they were efficient, especially when bad weather had tangled or flattened the grass. They were also light enough for women to use as well as men.

The earliest tool used for cutting hay was a reaping hook, so called because of its hooked shape. Using it was hard work, as the short handle meant you had to bend over all the time if you wanted to cut the stems close to the ground.

The next tool to be developed was the scythe, and a skilled man could mow three quarters of an acre with it in a day.

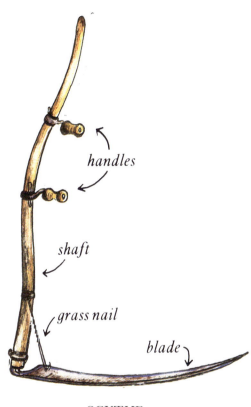

handles

shaft

grass nail

blade

SCYTHE

Its long handle made mowing easier on the back, but it was too heavy for women to use, and was not so good for cutting flattened grass. The wooden shaft and adjustable handles were made of ash. If a naturally curved bough could not be found for the shaft a straight piece was heated in steam and then bent. The length of the blade had to be the same as the distance between its tip and the lower handle on the shaft, or the scythe would not swing smoothly. A piece of wire called a grass nail was fixed between the shaft and the blade to stop the scythe from getting choked with mown grass.

whet stone

leather
belt
pouch

Blades were easily blunted by the tough stems of grass, so every mower carried a piece of sandstone called a 'whet stone' in a special pouch in his belt. He sharpened the blade by rubbing the stone along both sides, always from the shaft to the tip. A blunt scythe made a hard job harder, so the mower took good care of his whet stone.

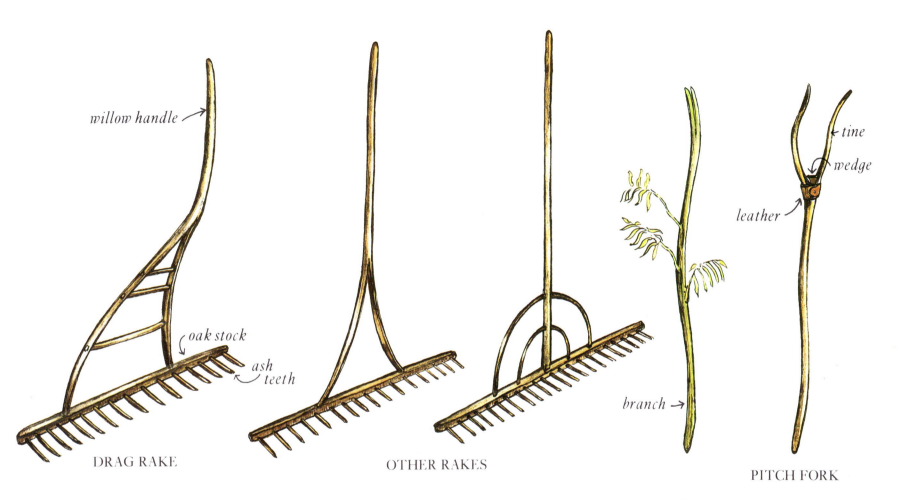

willow handle

oak stock

ash teeth

DRAG RAKE

OTHER RAKES

branch

tine

wedge

leather

PITCH FORK

When all the grass had been cut it had to be first turned, then raked into rows, where the wind and the sun would dry it. The design of the rakes varied in different parts of the country. One of the most popular, the drag rake, had a willow wood handle, a cross bar, or stock made of oak, and strong ash teeth which were fitted firmly into the stock. The drag rake was particularly good at raking up loose hay.

Another important tool was the pitchfork, used for tossing hay into the cart. To make it the farmer would choose a willow bough of the right length and thickness, then split it at one end. He would then bind the bottom of the split with leather to stop it splitting further and wedge a piece of wood between the two tines to keep them apart.

BUILDING
A HAYCOCK

SECTION THROUGH
HAYCOCK

pole

If his hay was to dry out properly on the ground, the farmer needed very fine weather. In bad weather, he had to stack his hay in 'haycocks', which could be anything from five to twelve feet high. If the haycock was a tall one, it would be built round a pole to keep it steady. When the haycock was as high as he wanted it, the farmer, would, as he said, 'comb the hay'. This meant he brushed the surface so that the stems hung downwards and overlapped slightly at the bottom. In this way he made sure that the rain ran off the surface of the haycock straight onto the ground, and the hay inside remained dry. Haycocks could be left in the field until the hay inside was thoroughly dry.

ladder

top rave   sideboard

hoop

head board

bed bolster

bond

shaft pin

axle bed

hub

shaft

axle

iron pin

dog stick
to stop cart
rolling back

felloes

spoke

PARTS OF A WAGON

PARTS OF A WHEEL

The horse-drawn wagon that carted the hay from the field was the farmer's most valuable possession.

The wagon was built in two separate parts: the body, which carried the load, and the undercarriage on which the body rested and to which the wheels were attached. The undercarriage was made in two parts called the forecarriage and the hindcarriage. The forecarriage swivelled freely on a long pin that was driven into a strong wooden support bar called a bed bolster; without the swivel action the wagon would have been impossible to manoeuvre through narrow gates and round corners. The wheelwright who made the wheels had to be a skilful craftsman and know his wood. He used elm for the hub, oak for the spokes and ash or beech for the rim. The rim was made by joining segments of the wood, called felloes, together. When the wheel was finished an iron hoop was heated at the forge, wrapped round the rim while still hot and then plunged into water. The cooling water shrank the iron tightly onto the wood.

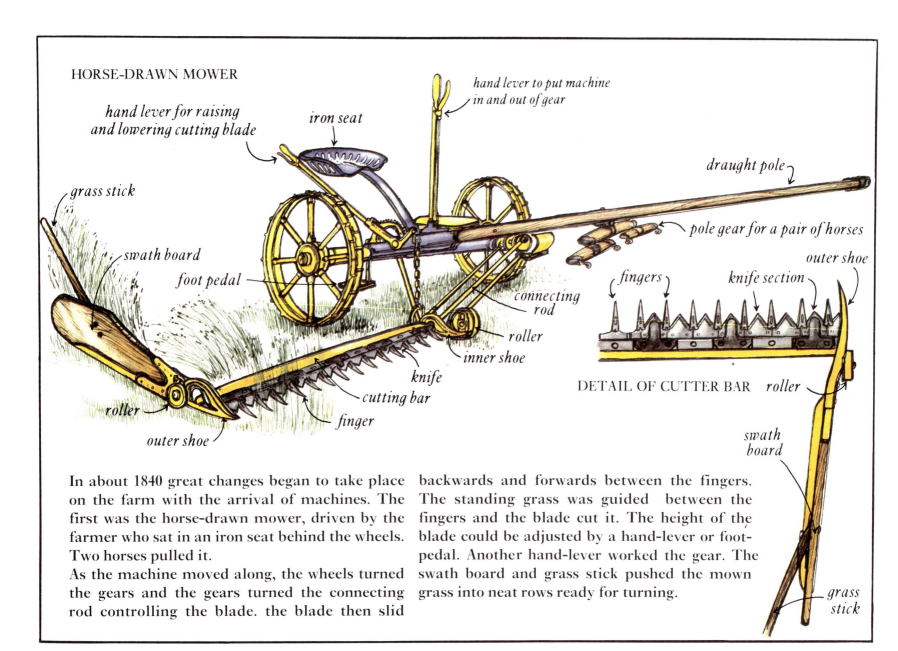

HORSE-DRAWN MOWER

hand lever for raising
and lowering cutting blade

iron seat

hand lever to put machine
in and out of gear

draught pole

grass stick

pole gear for a pair of horses

swath board

outer shoe

foot pedal

fingers

knife section

connecting
rod

roller

inner shoe

knife

cutting bar

DETAIL OF CUTTER BAR   roller

finger

roller

outer shoe

swath
board

grass
stick

In about 1840 great changes began to take place on the farm with the arrival of machines. The first was the horse-drawn mower, driven by the farmer who sat in an iron seat behind the wheels. Two horses pulled it.

As the machine moved along, the wheels turned the gears and the gears turned the connecting rod controlling the blade. the blade then slid backwards and forwards between the fingers. The standing grass was guided between the fingers and the blade cut it. The height of the blade could be adjusted by a hand-lever or foot-pedal. Another hand-lever worked the gear. The swath board and grass stick pushed the mown grass into neat rows ready for turning.

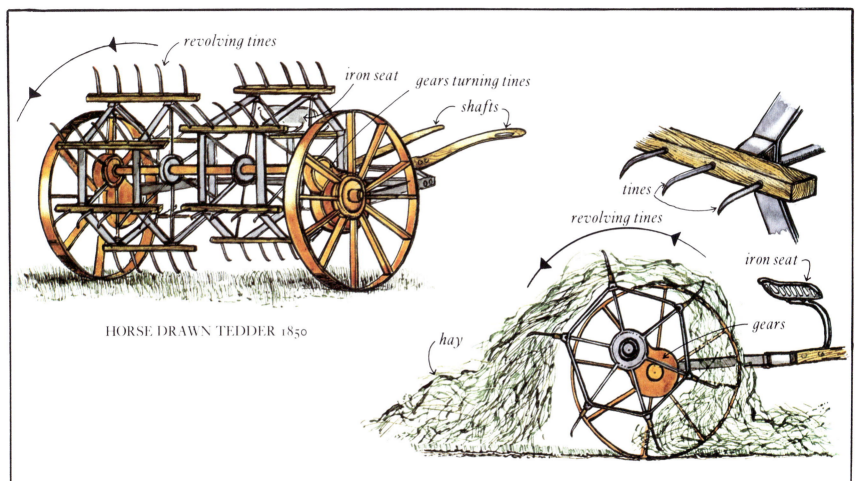

revolving tines

iron seat

gears turning tines

shafts

HORSE DRAWN TEDDER 1850

tines

revolving tines

iron seat

gears

hay

SECTION THROUGH TEDDER

The tedder, which is still used today, was the machine that did away with the job of turning the hay with a hand rake. It was pulled by a single horse, and its revolving tines tossed the hay into the air and turned it, so that the damp underside was facing the sun. The sharp, swiftly turning tines were very dangerous, and the farmer, perched above them on his iron seat, had to be very careful not to fall off.

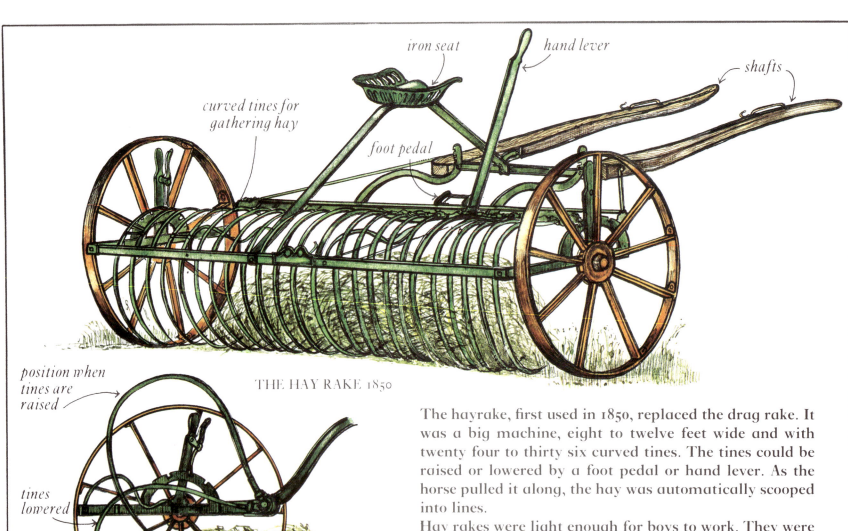

iron seat

hand lever

shafts

curved tines for
gathering hay

foot pedal

THE HAY RAKE 1850

position when
tines are
raised

tines
lowered

The hayrake, first used in 1850, replaced the drag rake. It was a big machine, eight to twelve feet wide and with twenty four to thirty six curved tines. The tines could be raised or lowered by a foot pedal or hand lever. As the horse pulled it along, the hay was automatically scooped into lines.

Hay rakes were light enough for boys to work. They were very popular, and when the tractor replaced the horse on the farm the shafts were adapted so the farmer could hitch them to his new machine.

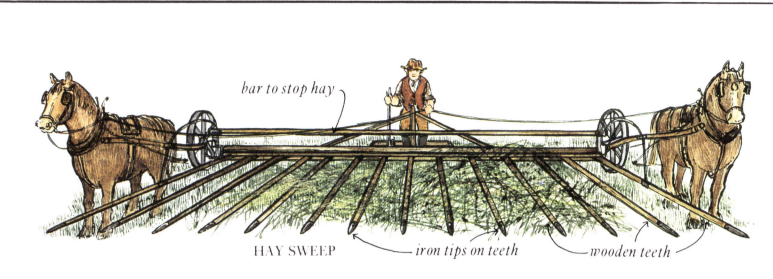

bar to stop hay

HAY SWEEP   *iron tips on teeth*   *wooden teeth*

The haysweep was designed for the wide, flat prairie lands of America where its great width made light work of the job of carting hay. It was never really successful in Britain, because, except in a few places like Essex and East Anglia, the fields were either too small or too hilly for it.

The wide framework was balanced on two wheels set at either end, behind the horses. The driver, perched on an iron seat behind the framework, raised and lowered the long wooden, metal-tipped teeth by pulling a hand lever.

hand lever to raise and lower teeth

seat

hay being swept up

SECTION THROUGH HAY SWEEP   *wooden teeth*

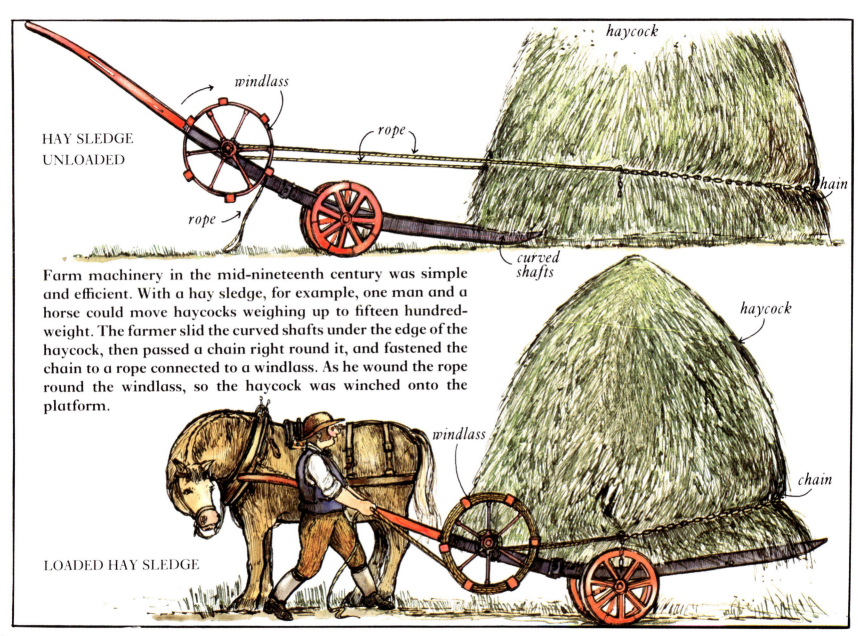

haycock

windlass

HAY SLEDGE
UNLOADED

rope

rope

chain

curved shafts

Farm machinery in the mid-nineteenth century was simple and efficient. With a hay sledge, for example, one man and a horse could move haycocks weighing up to fifteen hundred-weight. The farmer slid the curved shafts under the edge of the haycock, then passed a chain right round it, and fastened the chain to a rope connected to a windlass. As he wound the rope round the windlass, so the haycock was winched onto the platform.

haycock

windlass

chain

LOADED HAY SLEDGE

Sometimes the hillsides where the hay was cut were too steep to use either the hay sweep or the sledge. Then the haycocks were carried to a waiting cart by two men who slid eight foot poles set three feet apart under the haycock, balanced it carefully and then lifted poles and hay together.

In Derbyshire and other hilly counties in the north of England this method of carrying hay was called 'stanging', because 'stang' was the local name for a pole.

As the hay came into the yard it was built into a stack; a job that was done by hand, with pitch forks. Then, about a hundred years ago, someone invented the horse-powered elevator.

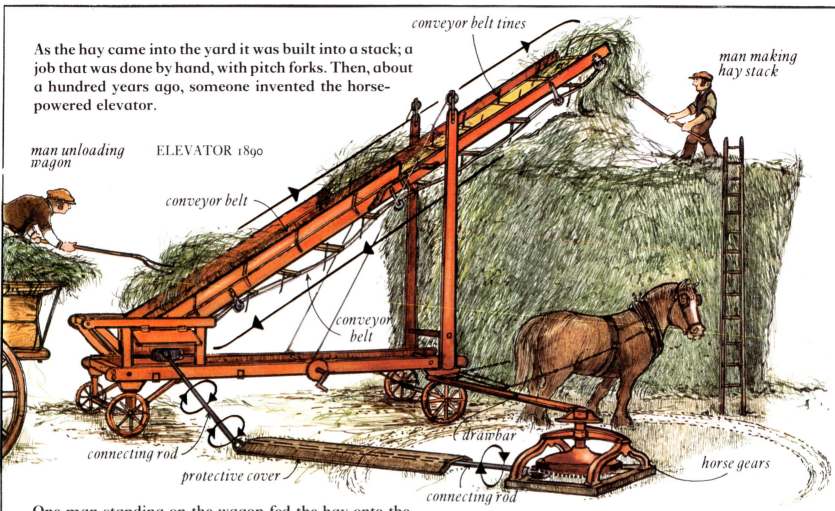

conveyor belt tines

man making hay stack

man unloading wagon

ELEVATOR 1890

conveyor belt

conveyor belt

connecting rod

protective cover

drawbar

connecting rod

horse gears

One man standing on the wagon fed the hay onto the moving belt which carried it to the top of the stack, where another man with a pitchfork spread it evenly. The patient horse walked round in a circle pulling a 'drawbar', one end of which was connected to the 'horse gears', as they were called. The gears turned the connecting rods, which were linked to the belt. The connecting rods were protected from the horse's hooves by a wooden cover.

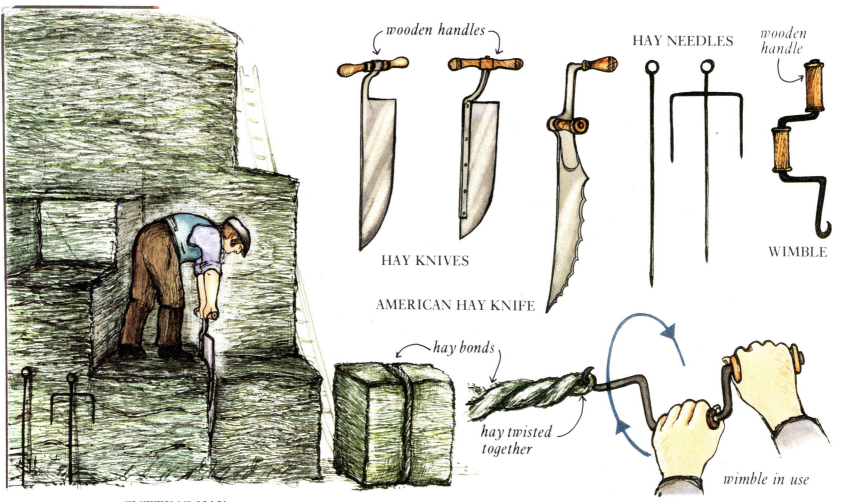

wooden handles

HAY NEEDLES

wooden handle

HAY KNIVES

AMERICAN HAY KNIFE

WIMBLE

hay bonds

hay twisted together

wimble in use

CUTTING HAY

If a farmer was lucky enough to have more hay than he needed, he could sell some and make a little money. To prepare the hay for sale he would first cut a block of about 56 lbs called a 'truss', then make sure the truss held together by pushing an iron hay needle firmly through it. Finally he made a rope, called a hay bond, with which he bound the truss. The ropes were made by twisting lengths of hay together with a 'wimble'.

HAYMAKING IN 1850

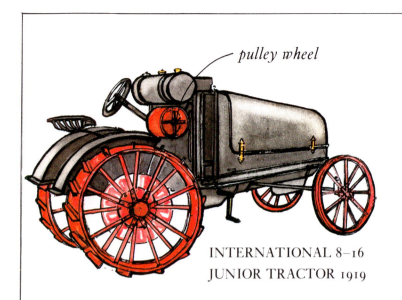

pulley wheel

INTERNATIONAL 8–16
JUNIOR TRACTOR 1919

FORDSON STANDARD TRACTOR 1919

In the early years of this century people got used to the extraordinary sight of motor cars on the roads, and the farmer began to use the tractor, a 'car' designed for jobs on the farm. The first tractors were very simple affairs with huge iron wheels and 25 horse power engines that ran on paraffin. Most of them were really like extra powerful horses, and only did 'pulling' work, but one or two had a pulley wheel that could drive a stationary machine like the baler on page 24.

FIAT MODEL 702

AUSTIN TRACTOR 1919

1946 was the year in which the first truly modern tractor, designed by Henry Ferguson, appeared. His tractor was *new* in two ways; it had a power-take-off shaft and a hydraulic lift system, both of which are important to tractor design today.

The power-take-off shaft *drove* implements, by transferring power to them directly from the tractor's engine. The tractor was no longer just a strong horse, but powered the machines attached to it, which up till then had needed forward motion to make them work.

lift rod

power-take-off shaft

lift arms

The hydraulic lift, fitted at the back of the tractor, made it possible to suspend machines such as mowers on movable arms with independent power. This gave the driver greater control than he had when a machine was simply pulled.

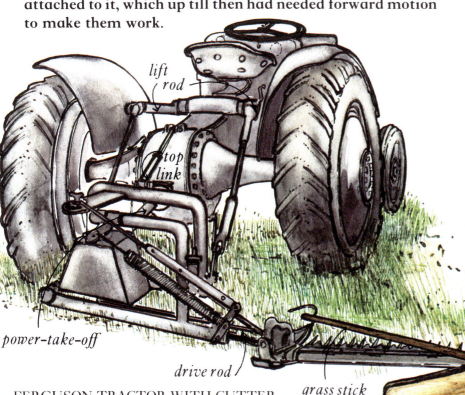

lift rod

top link

power-take-off

drive rod

grass stick

knife section

shoe

swath board

FERGUSON TRACTOR WITH CUTTER

hay fed into hopper

drive belt to tractor pulley

hay needles

finished bales coming out of baler

wire bonds

flywheel

hayneedle
wire bond

ram compressing hay

wire bond

finished bale

groove in needle

IRON NEEDLE

STATIONARY BALER

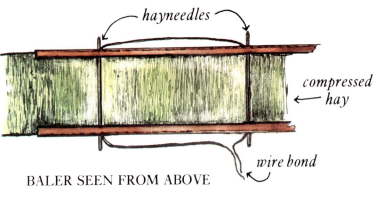

hayneedles

compressed hay

wire bond

BALER SEEN FROM ABOVE

Cutting hay from a stack and baling it was a slow job. The stationary baler, driven by the pulley on a tractor, meant the hay could be baled before stacking. The loose hay was fed into the hopper and compressed by the ram. Two men worked on the baling together. The first thrust an iron needle through the hay and pushed a length of wire through the groove in the needle to his companion. He pushed the wire back through the next needle. The first man secured the bale by twisting the ends together. These bales weighed up to 2 cwt, much heavier than today's bales.

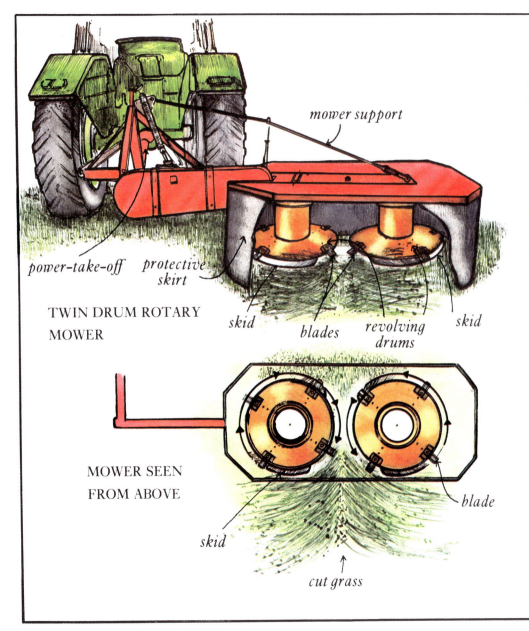

mower support

power-take-off

protective skirt

skid

blades

revolving drums

skid

TWIN DRUM ROTARY MOWER

MOWER SEEN FROM ABOVE

skid

blade

cut grass

The modern twin drum rotary mower is suspended behind the tractor on the hydraulic lift system and driven by the power-take-off shaft. The twin drums contra-rotate (go round in opposite directions) at high speed, and as the sharp revolving blades in each drum cut the grass, it falls in a neat swath between the drums. Attached to the bottom of each drum is a metal plate called a skid, which stops the blades touching the ground. A plastic skirt runs round the machine to stop stones flying up. This splendidly practical machine has few moving parts to go wrong, and blades that are easy to replace.

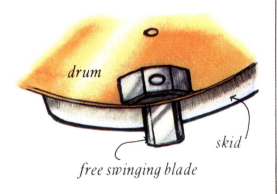

drum

skid

free swinging blade

DETAIL OF BLADE

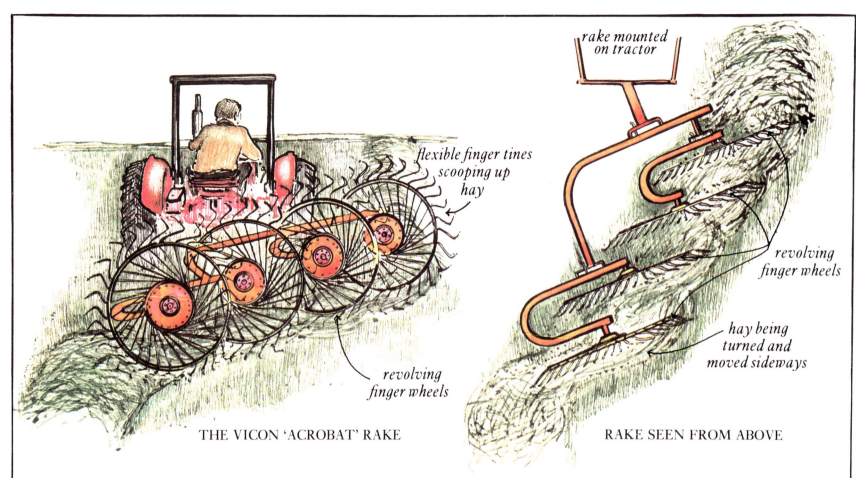

flexible finger tines
scooping up
hay

rake mounted
on tractor

revolving
finger wheels

revolving
finger wheels

hay being
turned and
moved sideways

THE VICON 'ACROBAT' RAKE

RAKE SEEN FROM ABOVE

This rake, or swath turner, one of the most popular machines on the farm, was designed about thirty years ago and nicknamed 'acrobat' because it was so springy. The four wheels, each fitted with a number of flexible wire finger tines, are staggered one behind the other at an angle giving the rake a wide reach. The wheels are turned as the tractor moves forward and the finger tines, in turn, make contact with the ground and rotate them. By adjusting the wheels this adaptable machine can be made either to gather or scatter hay according to what the farmer wants.

bale coming off
back of baler

bale density
adjustment

ram

flywheel

tail gate

*drive to
power-take-off*

bale chamber

draw bar

PICK-UP BALER

pick up tines

bale

hay in windrow

BALER TWINE

*hay bales stacked up ready for
collection*

Another modern version of an older machine is the pick-up baler. Today, instead of the hay being brought to the stationary baler, the baler goes to the hay, drawn by the tractor and driven from the power-take-off shaft. As the baler moves along the windrows, the pick-up tines gather the hay into the machine, where it is packed tight by a ram moving at the rate of 80 strokes a minute. The compressed hay is tied with baler twine from a container in the machine. These days polypropylene twine is most common – it is strong and does not rot. The finished bale is pushed through the tail gate and drops to the ground. When all the hay has been baled, the farmer stacks the bales for carting.

A pick-up baler makes a bale every 6 seconds. The farmer can adjust the ram to make tightly or more loosely packed bales, and he can vary the length of the bale – the shortest being 12 inches, the longest 50. Today's bales weigh about 56 lbs.

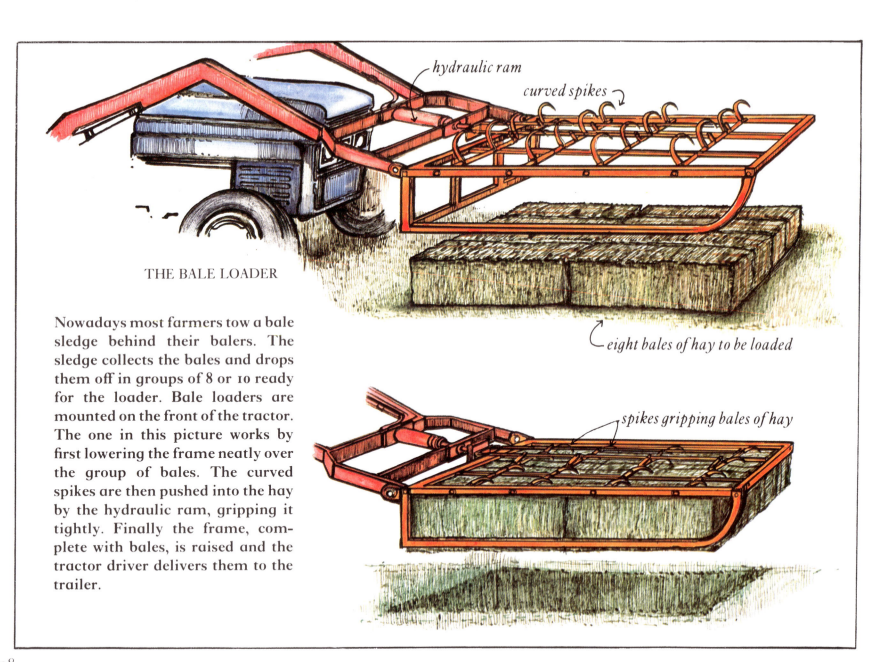

hydraulic ram

curved spikes

THE BALE LOADER

eight bales of hay to be loaded

spikes gripping bales of hay

Nowadays most farmers tow a bale sledge behind their balers. The sledge collects the bales and drops them off in groups of 8 or 10 ready for the loader. Bale loaders are mounted on the front of the tractor. The one in this picture works by first lowering the frame neatly over the group of bales. The curved spikes are then pushed into the hay by the hydraulic ram, gripping it tightly. Finally the frame, complete with bales, is raised and the tractor driver delivers them to the trailer.

When the bales of hay arrive at the stackyard they are lifted mechanically onto the stack by an elevator. No more back-breaking pitch-fork work!

Elevators have changed, too. They don't rely on the horse for power, but have their own petrol or electrically powered engines with an hydraulic ram to control the height.

From cutting to stacking, modern machinery helps the farmer to store his hay safely for the winter faster and with far less help than a hundred or so years ago.

THE ELEVATOR

conveyor belt

hydraulic ram

INDEX